For Nicola - A. T. B.
For Indiah and Lanta - A. A.

Let's Swim and Dive

Please visit our web site at: www.garethstevens.com
For a free color catalog describing Gareth Stevens' list of high-quality books and multimedia programs, call 1-800-542-2595 (USA) or 1-800-461-9120 (Canada).
Gareth Stevens Publishing's Fax: (414) 332-3567.

Library of Congress Cataloging-in-Publication Data

Nilsen, Anna, 1948-
Let's swim and dive / written by Anna Nilsen; illustrated by Anni Axworthy.
p. cm. – (Animal antics)
ISBN 0-8368-2913-1 (lib. bdg.)
1. Animals–Miscellanea–Juvenile literature. 2. Animal swimming–Juvenile literature.
[1. Animal swimming.] I. Axworthy, Anni, ill. II. Title.
QL49.N543 2001
590–dc21 2001020882

This North American edition first published in 2001 by
Gareth Stevens Publishing
A World Almanac Education Group Company
330 West Olive Street, Suite 100
Milwaukee, WI 53212 USA

Gareth Stevens editor: Dorothy L. Gibbs
Cover design: Tammy Gruenewald

Printed in the United States of America

1 2 3 4 5 6 7 8 9 05 04 03 02 01

Animal Antics

Let's Swim and Dive

Written by
Anna Nilsen

Illustrated by
Anni Axworthy

Gareth Stevens Publishing
A WORLD ALMANAC EDUCATION GROUP COMPANY

Whales squoosh
up water
before they

swim
and
dive.

Octopuses wave their arms and dance when they

swim
and
dive.

Hungry alligators whip their tails from side to side as they

swim
and
dive.

Duck-billed platypuses close their eyes while they

swim

and

dive.

In the deepest, darkest seas,

little fish gleam and glow as they

swim
and
dive.

Polar bears
must paddle
with their
paws to

swim

and

dive.

To build
their dams
and lodges,
beavers
carry logs
while they

swim

and

dive.

Heavy hippos often tiptoe down the river when they

swim
and
dive.

Even girls and boys like you sometimes

swim
and
dive.